THE
GOOSE MAN
The Story of Konrad Lorenz

ELAINE GREENSTEIN

CLARION BOOKS
HOUGHTON MIFFLIN HARCOURT
BOSTON • NEW YORK • 2009

Clarion Books
215 Park Avenue South
New York, NY 10003
Copyright © 2009 by Elaine Greenstein

The illustrations were executed in gouache, ink, and colored pencil.
The text was set in Bodoni.
Book design by Janet Pedersen.

Special thanks to Tarquin Mittermayr, archivist at the Konrad Lorenz Institute, for all the help with the
photographs, and especially for finding out about the chandelier in the hallway.

For information about permission to reproduce selections from this book, write to Permissions,
Houghton Mifflin Harcourt Publishing Company, 215 Park Avenue South, New York, NY 10003.

Clarion Books is an imprint of Houghton Mifflin Harcourt Publishing Company.

www.clarionbooks.com

Printed in Singapore

Library of Congress Cataloging-in-Publication Data
Greenstein, Elaine.
The goose man : the story of Konrad Lorenz / Elaine Greenstein.
p. cm.
Includes bibliographical references.
ISBN 978-0-547-08459-6
1. Lorenz, Konrad, 1903–1989–Juvenile literature. 2. Ethologists–Austria–Biography–Juvenile literature.
3. Geese–Behavior–Juvenile literature. I. Title.
QL31.L76G74 2009
590.92–dc22
[B]

2008044618

TWP 10 9 8 7 6 5 4 3 2 1

For Joann Hill

When Konrad Lorenz was six years old, he and his friend Gretl practiced being ducks. They made duck noises, they walked like ducks, they splashed like ducks—they even tried to fly like ducks.

Each of them was given a newly hatched duck. The ducklings didn't
seem to notice that the children weren't ducks. The ducklings followed
them wherever they went and tried to do whatever they did.

The duck was the first of Konrad's many pets. Then came fish, dogs, insects, an attic full of birds, a crocodile, and various monkeys. His parents shook their heads and sighed. They loved their son, and their son loved animals so much that they let him turn their house into a zoo. They ignored the mess the pets made and warned their guests to hide anything shiny . . . or risk having a monkey snatch it.

Every March, hundreds of geese flew north over Konrad's house
on their way to their summer homes.

Every October, hundreds of geese flew south over Konrad's house
on their way toward warmer places.

Konrad loved how they honked to each other.

He loved how they flapped their wings.

He loved to watch them flying.

He wanted to fly with them.

Konrad watched the geese.

As Konrad grew up, he kept watching wildlife.

His friend Gretl became his wife.

He became a doctor, but he loved learning about animals more than he loved practicing medicine.

He became a scientist and studied how animals behave.

When he was thirty years old, he devoted most of his time to studying geese.

Konrad wanted to learn more about what happens when a goose is born. A friend helped him get some wild goose eggs. Konrad placed the eggs in the nest of a pet goose that lived in his garden.

Konrad checked the nest every morning.
In about four weeks, he started to hear chirping
inside the shells: *vee-vee-vee.*

One day soon after, the
eggs began to hatch.
The chirping inside the
eggs became louder.
The foster mother chirped
back at the eggs.
Konrad chirped, too.

Soon a crack appeared, and then another.
Soon one little beak made its way
through a shell, then another.
Konrad watched. He watched for hours.

Shortly after one of the goslings hatched, Konrad picked her up. Looking right into Konrad's eyes, the little gosling began to chirp.

He had watched enough geese to know that the sound meant "I'm lost," and he knew what to chirp back to calm her.

Then the gosling stretched her neck and called *vee-vee-vee,* and he knew that this was a happy greeting. Each time the gosling cried out, Konrad made comforting sounds.

It was time for Konrad to head home. He placed the little hatchling with the rest of the brood and started to walk away. The gosling chirped and ran after him. At that moment, Konrad knew: The baby goose had decided that *he* was her parent.

He tried one more time to put her near her foster mother, but she ran
back to him.
Konrad tucked the gosling into his shirt and took her home.
He named her Martina.

Every morning, Konrad read in the garden while Martina nibbled on grass.

Every afternoon, Konrad and Martina swam in the river.

He often tucked Martina inside his shirt while he worked at his desk.

Every evening, Martina waited on Konrad's doorstep.

When the door opened, she ran in and up the stairs to the bedroom.

Every night, Martina slept in the bedroom.

Every morning, she flew out the window.

Gretl just shook her head and put up with the mess the goose made.

Even though Martina was a pet goose,
she had been born knowing some wild things.
Martina honked to the geese flying overhead in March.
Martina honked to the geese flying overhead in October.

Vronk vronk vronk!

Konrad learned what Martina's different honks meant.
Carun-kk meant "Rain coming."
Honk-honkkkk meant "Time to eat."
Vronk-vronk vronk meant "Good place to rest."

17

In March, Martina and Konrad did their usual honking to the geese that were flying over the garden.

One day a goose chattered back and landed.

That evening, Konrad was surprised to find *two* geese on his doorstep.

Konrad named the new goose Martin.
Martin had never been in a house, but he followed right
behind Martina.

When the door closed, the sound frightened him. He flew straight up
to the chandelier. Down fell some of the glass dangles and also a few
feathers. Konrad and Martina had to do a lot of honking to get Martin
back to the ground.
Next morning, two geese flew out of the bedroom window.

Now, every morning, two geese nibbled on grass in the garden.
Now, every afternoon, two geese followed Konrad to the river
for their daily swim.

Now, every evening, two geese waited on the doorstep.
Now, two geese chattered with Konrad to the geese flying overhead.

But one evening, Konrad was surprised to find no geese on his doorstep.

No geese at the back door.

No geese in the garden.

Gretl helped him look for Martina and Martin, but Konrad never saw them again. He decided that they had flown away with their wild relatives.

Konrad raised many geese.

He made sure he was there to bond with them when they hatched.

Konrad sat near many nests.

The annoyed parent geese hissed at him,

but Konrad knew what to hiss back.

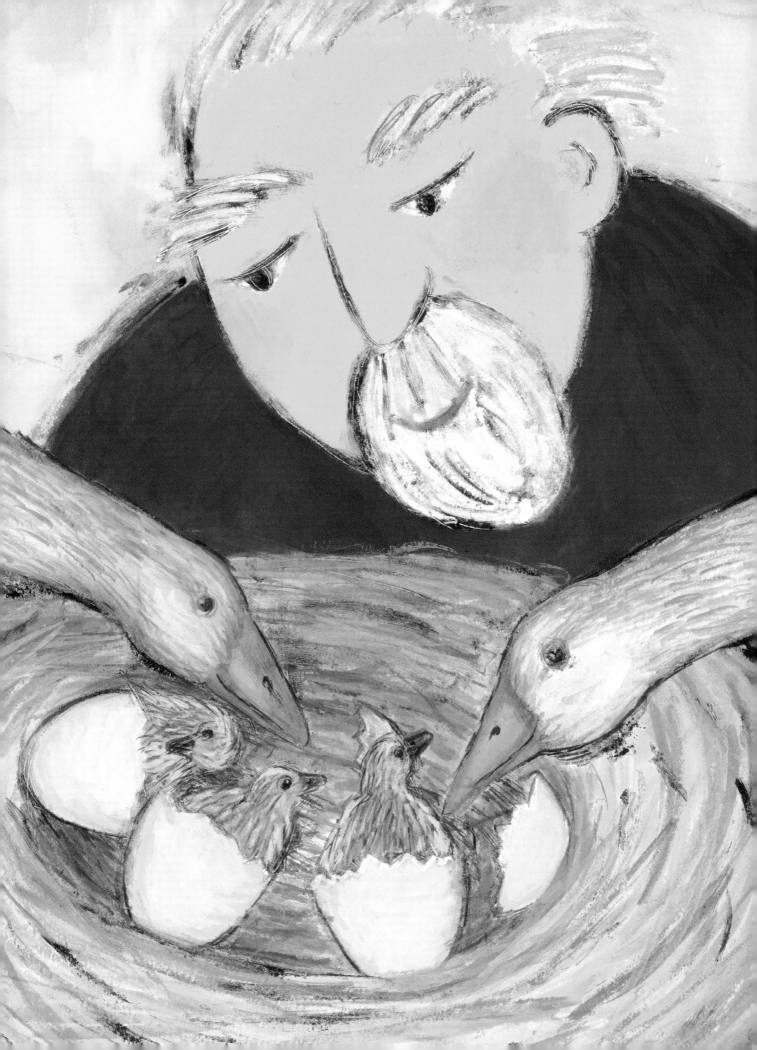

Geese followed Konrad everywhere.
When they were little balls of fluff,
when the little geese began to grow feathers,
and when they started to look like grown-up geese,
they followed Konrad everywhere.

Martina and Martin were the only geese that ever slept in Konrad's bedroom every night.
Sometimes Konrad slept outside with the geese to learn what they did at night, but he wasn't a goose, and he missed his bed.

Konrad raised so many geese that he couldn't watch them all. He hired
helpers to follow the geese and write down whatever they did.

Every day, Konrad swam in the river.
Geese always swam with him.
Konrad always watched the geese.

Konrad kept watching the geese.

He learned that geese are born knowing wild things—

like goose language, and what to do every day and when—

even if they are not wild.

He learned that some geese attach themselves to the first thing

they see, even if it's not a parent.

When Konrad was an old man, he won a big, important prize

for the new things he had learned.

AUTHOR'S NOTE

Konrad Lorenz (1903–1989) was born in Austria. His father wanted him to be a doctor, but Konrad was fascinated with wildlife. Although he became a doctor, after a few years he left his practice to study what he loved. Lorenz founded a science of animal behavior called ethology. Instinct is behavior that appears at birth and does not need to be learned. Some of a goose's instinctual behaviors are flying, language, swimming, courtship, and imprinting, which means attaching itself to the first thing it sees. Konrad Lorenz's work on imprinting is one of the reasons he won the Nobel Prize in 1973. This story is based on the research he did beginning in 1934, when he focused on geese.

SOME BOOKS ABOUT KONRAD LORENZ AND HIS WORK WITH GEESE

Burkhardt, Richard. *Patterns of Behavior: Konrad Lorenz, Niko Tinbergen, and the Founding of Ethology*. Chicago: University of Chicago Press, 2005.

Evan, Richard I. *Konrad Lorenz: The Man and His Ideas*. New York: Harcourt Brace Jovanovich, 1975.

Lorenz, Konrad. *King Solomon's Ring: New Light on Animal Ways*. New York: T. Y. Crowell, 1952.

——. *The Year of the Greylag Goose*. New York: Harcourt Brace Jovanovich, 1979.

Lorenz, Konrad, and Sybille Kalas. *The Goose Family Book*. New York: Michael Neugebauer Books/North South, 2000.

Lorenz, Konrad, Michael Martys, and Angelika Tipler. *Here I Am—Where Are You? The Behavior of the Greylag Goose*. New York: Harcourt Brace, 1988.

Nisbett, Alec. *Konrad Lorenz*. New York: Harcourt Brace Jovanovich, 1977.